THE WORLD'S GREATEST UFO MYSTERIES

PHOENIX W. THOMPSON

Copyright © 2022 Phoenix W. Thompson

All rights reserved.

ISBN: 9798847705691

DEDICATION

For mother, at her request of a paperback version.

CONTENTS

IMPORTANT NOTE

The events contained in this book are accounts of unusual encounters as reported at the time.
The author makes no definitive claim that what was reported is extra-terrestrial in nature, simply that something happened on these occasions involving unidentified flying objects.
The following incidents remain largely unexplained…

INTRODUCTION

Throughout history human beings have looked to the stars for answers to the unexplained. From the question of who we are, to the question of where we came from, many believe the truth will be found somewhere in the vastness of space as opposed to here on Earth.

No question has been more prominent over the last century than the question of whether we are alone in the universe. However, even if it were discovered that humanity isn't the only intelligent civilization to have evolved, what are the chances that we have been visited?

Some believe that the sheer distances needed for interstellar travel would make it almost impossible for an extra-terrestrial race to visit us, even at the speed of light, while others remain convinced that not only is it possible, but that it may already have happened.

The existence of unidentified aerial phenomenon is not just a modern development. There have been reported sightings of unidentified flying objects for almost as long as human beings have walked the earth.

There have been depictions of strange objects in the sky - dating back over ten-thousand years - found in recently discovered stone carvings in India, which look remarkably similar to the typical "flying saucer" that you might find in a Hollywood movie.

In 214 BC, Titus Livius Patavinus, and ancient historian, wrote an account of "phantom ships seen gleaming in the sky", which is widely regarded as one of the earliest written records of an unidentified flying object.

During the 15th century, many renaissance paintings depicted unusual objects in the sky. One in particular, Madonna with Saint Giovanni by Domenico Ghirlandaio, shows a strange object in the clouds, that shares many similarities to many 20th century eyewitness accounts of UFOs.

It could be that these are nothing more than mere coincidences, but what if they weren't? What if we had been visited by extra-terrestrials throughout history? Some UFO enthusiasts believe that this may account for many of the religious and historical oddities that humanity has encountered throughout our existence.

Could it be that the bright star leading the wise men to the birth of Jesus Christ could have been an extra-terrestrial craft?

What about the stone tablet handed down to Moses containing the ten commandments? Could this have been the work of extra-terrestrials giving humanity insights into how they should live in order to develop a peaceful society?

Some believe that is exactly what happened.

While it may seem unbelievable, you have to ask yourself, is it not just as believable as the premise that there's an all-powerful being who created the world and everything in it?

Some would argue it's more likely that extra-terrestrials are responsible than a God of some kind.

Whatever you believe, there's no escaping the fact that it would be a mathematical improbability for Earth to be the only planet in the universe to harbour intelligent life.

There are more stars in our galaxy alone than there are grains of sand on our entire planet, and more galaxies in the universe than could be counted in a lifetime.

There is a strong belief throughout UFO enthusiasts that not only has the earth been visited by other intelligent civilizations, but that the governments of the world are well aware of their existence and even have extra-terrestrial technology in their possession, but refuse to disclose the truth to the public for fear of mass panic and civil unrest.

Could you imagine waking up to the news tomorrow and hearing that extra-terrestrials exist and the government has known about it this entire time? Many believe it would be catastrophic for humanity as a whole.

Religious organizations would crumble, the economy would collapse, people would be scared and mass hysteria would ensue. They agree that if the government do know something, they are right to keep it from us, as it isn't in our best interests to know.

Others, however, believe that we have been drip-fed information over the years through depictions of extra-terrestrials in popular culture, such as television shows, movies, books and games, in an attempt to condition the human race so that society doesn't break down should the truth come out.

Then there's those that believe the public has a right to

know, no matter the consequences. Something so significant, so life-changing, should be made public to reassure the people that the government are prepared to defend us from any threat, terrestrial and otherwise. This book is not intended to convince you one way or the other, but simply to detail some of the most significant events in UFO history and allow you to form your own opinion and follow the rabbit hole deeper, should you choose to do so.

These are the world's greatest UFO mysteries…

AURORA, TEXAS
(15TH APRIL, 1897)

History is littered with some sensational stories relating to the UFO phenomenon, and the many theories put forward by experts in the field have often struggled to produce any conclusive proof one way or the other. Some explanations seem to be so far-fetched they could be pulled straight out of a Hollywood blockbuster, while others seem too simple to account for the incredible stories put forward by eye-witnesses.

Since air travel has become a daily occurrence over the last century, it stands to reason that many UFO sightings could be explained away as the simple misidentification of a terrestrial aircraft, but what about events that took place before we took to the skies?

In the small town of Aurora. Texas, an event was reported that continues to baffle the UFO community to this day, but what makes the Aurora incident so significant, is that it remains possible that some actual physical evidence of an extra-terrestrial encounter may still exist just waiting to be discovered.

Between 1896 and 1897 there were multiple stories being reported in the news relating to unknown aerial phenomena. There were numerous accounts of "airships" being made public, beginning in California and travelling east.

These were described as being metallic, cigar-shaped objects that were slow-moving and carried incredibly bright spotlights.

While the first airship was built in 1852, they were far from a common sight. Some people have claimed that these objects were "phantom airships", while others maintain that they were a form of Zeppelin.

What's interesting to note, however, is that the Zeppelin wasn't patented in the United States until a few years later, in 1899, and didn't start flying commercially until 1899.

The amount of people witnessing these objects was so unusual, that the media began reporting on them frequently, though none of those reports were as important to UFO researchers as a small newspaper article written by S.E. Haydon and published in The Dallas Morning News on April 19th, 1897.

The article reported that on 15th April, 1897, an "airship" came sailing down from the sky heading north, and travelling at between ten and twelve miles per hour.

There were many witnesses to the event, who watched on as the ship seemed to experience some mechanical difficulties which caused it to collide with a windmill on the property of the local Judge, J.S. Proctor. According to reports, this resulted in a huge explosion that caused debris to scatter over several acres. When the wreckage was investigated by the locals, they described the debris as a strange, light metal that none of them had encountered before.

The article stated is appeared to be a "mixture of aluminium and silver, and it must have weighed several tons". The craft had been so severely damaged in the explosion that no conclusions could be drawn relating to its construction.

For a small town like Aurora, this would have been a spectacular sight to behold, but the crash was only the start of this otherworldly encounter, as searching through the rubble led to the grim discovery of a body amongst the wreckage.

The body in question was described in the article as being the only one on board and assumed it to belong to the pilot. The next part of the article thrust this incident into the spotlight of many UFO researchers; "while his remains are badly disfigured, enough of the original has been picked up to show that he was not an inhabitant of this world."

It could well be the case that the crash had damaged the body so badly that it could have been a human pilot simply misidentified, but the article also claims that some strange papers were found on the body that were written in some "unknown hieroglyphics" that could not be deciphered. This raises the question as to how the papers could survive if the body was too damaged to identify as human.

According to witnesses, people descended on the town to view the wreck and gather specimens of the material, while the townsfolk held a funeral for the "pilot", giving it a Christian burial in the nearby cemetery complete with a crude headstone etched with the image of the craft as a grave marker. When people had finished taking specimens from the wreckage, the remaining debris was dumped into an old well. Some sceptics argue that since the article was written by an Aurora local, it must have been a joke piece. However, in the years that followed, numerous witness have come forward to corroborate the account of The Dallas Morning News, that have added credibility to the account. Some stated they saw the craft, while others claim to have seen the crash itself, and the dead body of the "pilot".

The incident seemed to have been forgotten for many years until there was a strange claim in 1945 from Brawley Oates, who had purchased Judge Proctor's property. Mr. Oates claimed that he had found some strange debris in the well on his property, and that it had made him so ill he had been forced to cover it with cement to seal it away.

At the time, Mr. Oates' claims, and the crash itself, were as sensational a story as you would find in any newspaper at the time, but in the years that followed, it seemed to fade away from people's memories and almost disappeared into history until decades later when the story was rediscovered by UFO enthusiasts and it quickly became one of the most popular events to investigate.

One of the earliest investigations into the Aurora incident was that of Mr. Bill Case. Mr. Case was not only an aviation writer for the Dallas Times Herald, but also the Texas State Director of the Mutual UFO Network (MUFON). During his investigation, Mr. Case interviewed surviving witnesses and visited the cemetery.

Two new witnesses were identified during the investigation. The first, Mary Evans, claimed to have been fifteen at the time of the incident and explained how her parents had visited the crash site and witnessed the discovery of the body. She was forbidden from going by her father.

The second was Charlie Stephens who was ten years old at the time of the crash and stated that he had witnessed the airship heading north towards Aurora while trailing smoke. He wanted to go and see the crash site, but instead his father made him do his chores. However, his father did visit the site himself the following day.

In 1973, following his investigation, Mr. Case claimed to have discovered the gravestone that had been left as a marker of the body buried back in 1897. What remained was a small, half-broken rock with what Mr. Case believed to be part of a saucer-shaped etching. During his trip to the cemetery, he used a metal detector at the gravesite and it appeared to Mr. Case that there were large metal pieces buried in the ground. Mr. Case took his findings to city officials and asked for permission to exhume the grave. His request was denied, despite repeated attempts. When Mr. Case returned to the site, he claimed that the metal pieces were no longer present.

This has led to many believing that they were intentionally moved by someone in an attempt to hide evidence. Shortly after this, the etched gravestone went missing and the exact location of the grave was lost again.

Discussions surrounding the reported crash in Aurora seemed to die down once again until 1979, when Time Magazine interviewed Aurora resident, Etta Pegues, who claimed that the whole incident was fabricated by S.E. Hayden as an attempt to bring interest to Aurora, since the town had been dying prior to the events of

1897. She also claimed that Judge Proctor never owned a windmill, which doesn't align with the initial reports and led to many believing the whole event was an elaborate hoax, created to drum up business for a small town suffering economic decline.

In 2008, the television show, UFO Hunters, carried out their own investigation into the Aurora incident. The team contacted Tim Oates, the grandson of Brawley Oates, and current owner of Judge Proctor's property, to ask for permission to unseal the well on the property in an attempt to uncover parts of the wreckage. Samples of water were taken from the well which tested normal apart from high amounts of aluminium. Though disappointingly for the team, no debris from the wreckage was found in the well and it was stated that the previous owner of the property had removed large pieces of metal and somehow disposed of them. What the team did find near the well, were the remains of a windmill, which contradicts the statement made by Etta Pegues to Times Magazine around 30 years earlier that Judge Proctor never had a windmill on his property.

The team also examined the cemetery. Using old photographs, they were able to locate the site where the missing gravestone once stood. When the team scanned the area with ground-penetrating radar, they discovered an spot which appeared to show that something was in fact buried amongst the graves from the 1890s. However, once again they were disappointed as their request to exhume the site was denied by the cemetery association.

To this day, the cemetery at Aurora still bears a plaque that lists an alien as one of those interred there, which could add credibility to the tale. Likewise, it could also be nothing more than an attempt to capitalize on the unexplained events of 1897.

Whatever the case may be, what is clear is that there is at least some truth to the story of the Aurora crash. Something is undoubtedly buried in a small, unmarked area of the local cemetery. Whether that is the remains of an interstellar visitor who came to its end in a fiery explosion on Judge Proctor's farm, or whether there's a simple, terrestrial explanation, is up for debate. What is absolutely certain, however, is that the truth is buried six feet underground in the cemetery of the small unassuming town of Aurora, Texas…

ROSWELL, NEW MEXICO
(4TH JULY, 1947)

In June of 1947, the term "flying saucer" was coined by the press when private pilot Kenneth Arnold reported seeing nine metallic discs flying by Mount Rainier at speeds that he estimated were in excess of twelve hundred miles per hour.

This caused a flying saucer craze to sweep across the globe which resulted in a large amount of sightings being reported, most of which had simple explanations. However, none of those reports created as much buzz throughout the UFO community than the one reported the following month near the small town of Roswell in New Mexico.

There is some disagreement over when the events at Roswell first began to unfold, but it is agreed that it was either July 2nd or July 4th of 1947 with the latter being widely regarded as the correct date.

The story begins with a rancher by the name of William Brazel, also known as Mac. One night during a typical summer thunderstorm, he heard what sounded like an explosion. The following day, Mr. Brazel and his seven year old neighbour, Dee Proctor, rode out to check that the fences hadn't been damaged by the storm.

As they came to the field ahead, they noticed that the ground was littered with debris of some kind. The debris was shiny and metallic in appearance and stretched over a quarter of a mile out in front of them. Mr. Brazel picked up some of the debris and carried it back with him.

Mr. Brazel's daughter, Bessie, later stated during interviews that her father had showed her some of the debris he had collected that night, and she remembered it looking like a sort of aluminium foil.

After finishing his work for the day, Mr. Brazel took Dee Proctor home. Dee and his parents were Mr. Brazel's closest neighbours and lived roughly eight miles away. He took some of the debris he had found along with him and showed it to them. He tried to convince Dee's parents to return with him to the ranch so they could see the field for themselves.

According to Floyd Proctor, Dee's Father, the material Mr. Brazel showed him was unlike anything he had ever seen. It couldn't be cut with a knife and had designs on it that was like Chinese writing or hieroglyphics.

Loretta Proctor, Dee's Mother, corroborated her husband's account, but added that there was a light brown piece of debris that was about four inches long and very lightweight like balsa wood.

When Mr. Brazel told his uncle Hollis about what he had discovered, he suggested that his nephew report the debris as there had been recent sightings of flying saucers in the area.

On July 6th, Mac was heading to Roswell so stopped off at the Chavez County Sheriff's Office to make his report. Sheriff George Wilcox originally didn't think anything of Mr. Brazel's find until he handled some of the debris himself. At that point he telephoned the nearby Roswell Army Air Field and spoke to Major Jesse Marcel.

Major Marcel was the chief intelligence officer at the base and told Sheriff Wilcox that he would come and talk to Mr. Brazel himself. However, before he had the chance, Mr. Brazel had already spoken to the local radio station, KGFL, about the incident.

Major Marcel met Mr. Brazel at the Sheriff's office. Mr. Brazel explained everything to Major Marcel and showed him the debris he had taken to the Sheriff. Major Marcel reported back to his commanding officer, Colonel William Blanchard, who subsequently ordered Major Marcel to travel to Foster Ranch and investigate the debris field.

Major Marcel was accompanied by another Officer, Sheridan Cavitt, to the ranch with Mr. Brazel, but since it was too late to visit the field by the time they got there, they all stayed at Mr. Brazel's house and decided to head out at first light. Once he had taken them to the debris field, Mr. Brazel left Major Marcel and Officer Cavitt to conduct their investigation.

.Local radio reporter Frank Joyce had by this point informed his boss, Walt Whitmore, about the events that had been unfolding. It has been reported that Mr. Whitmore interviewed Mr. Brazel at his home on record, but the recording has never been heard by the public.

Once Mr. Whitmore and Mr. Brazel returned to the radio station, Mr. Whitmore called the base. At this point, things took a strange turn.

Mr. Brazel was picked up by the military and taken to

the base. It's not known why he was taken to the base, nor is it known what the purpose behind his stay was. What is known, is that on July 8th, Mr. Brazel was returned to Roswell where a press conference was held. After Mr. Brazel returned from the base, his story had changed from what he initially reported. During the press conference, he stated that the debris was discovered by his son in mid-June, and that he was so busy that he ignored it. It wasn't until July 4th, that he visited the field himself and stated that he had found tin foil, paper, wooden sticks, and grey rubber pieces. He did state that he had found weather balloons on many occasions, and that this was different. He was quoted as saying, "I am sure what I found was not any weather observation balloon, but if I find anything else beside a bomb, they are going to have a hard time getting me to say anything about it".

After the press conference, Mr. Brazel was escorted to the radio station where he started to tell Frank Joyce the story he had just told the other reporters. Mr. Joyce interrupted Mr. Brazel at one point asking him why his story was different to what he told him initially. Mr.

Brazel simply told him, "it'll go hard on me".

Mr. Brazel was returned to the base after his interview with Joyce, and later released. Following his time on the base, Mr. Brazel refused to discuss what he had found with those around him.

However, some of those closest to Mr. Brazel have since reported that he would complain about being treated harshly by the military, about how they wouldn't let him call his wife while he was there, and about how he had been made to swear never to discuss the debris field again.

In less than a year after his discovery, Mr. Brazel moved away from the area, eventually passing away in Tularosa in 1963.

While Mr. Brazel was being interviewed by Mr. Whitmore, and prior to his stay on the base, Major Jesse Marcel was searching through the debris field with Sheridan Cavitt.

Major Marcel recalled what he had witnessed in later interviews. He stated, "When we arrived at the crash site, it was amazing to see the vast amount of area it covered. It scattered over an area of about three-quarters of a mile long, and fairly wide, several hundred feet wide". He went on to say, "I don't know what it was, but it certainly wasn't anything built by us and it most certainly wasn't any weather balloon". What's worth noting, is that Major Marcel was the Chief Intelligence Officer at the base, which was home to the 509th Bomb Group. The only one in existence at the time. He was a pilot during World War Two, and was highly decorated, previously working on the nuclear project, Operation Crossroads, in 1946. Major Marcel was a man of impeccable character. It's also worth noting that all personnel on the base had high level security clearance due to working with the 509th.Bomb Group. This adds credibility to Major Marcel's testimony and makes it difficult to believe, given his credentials, that he would misidentify the wreckage.

When describing the debris, Major Marcel's account corroborates that which was initially reported by others. He stated that the material was thin like tin foil, but could not be cut or burned, and went even further stating, "We even tried making a dent in it with a sixteen-pound sledgehammer, and there was still no dent in it".

Officer Cavitt filled his jeep with some of the material and returned to the base, while Major Marcel took a small box of debris with him in his car, stopping off at his house to show it to his wife and son.

His son would later recall during interviews his father waking him up in the night to show him the find. His description of the debris aligns with the others stating that it was foil-like, very thin, and very tough.

When Major Marcel returned to the base the following day, he was instructed by Colonel Blanchard to fly with the debris on board a B-29 to Wright Field in Ohio, via Fort Worth in Texas.

Meanwhile, Colonel Blanchard issued an order to Colonel Walter Haut at Roswell, to write a press release that the Roswell Army Air Force had in its possession a crashed flying saucer.

Colonel Haut later stated that the saucer was taken to the 8th Air Force under the supervision of General Roger Ramey.

The headline in the Roswell Daily Record would

become one of the most famous. It read, "RAAF CAPTURES FLYING SAUCER ON RANCH IN ROSWELL REGION".

When Major Marcel made his way to Fort Worth, General Ramey, who was the Commander in Charge, took full control over the debris.

Some of the debris was taken into the office of General Ramey and photographed. Major Marcel was also photographed with the debris. However, shortly afterwards, Major Marcel was taken into another office by General Ramey and when they returned, the material spread out on the floor was different to what Major Marcel had escorted to the base.

Under orders from General Ramey, Major Marcel stated that the debris was from a weather balloon. When they were finished taking photographs, General Ramey ordered Major Marcel back to Roswell with a warning to not speak a word of what he had witnessed while on the base. Following the trip to Fort Worth, it was reported by the press that General Ramey was the one who had identified the debris as parts of a weather balloon.

Many years later, the Chief of Staff at the 8th Air Force, Brigadier General Thomas DuBose would break his silence on the subject and state, "It was a cover story. The whole balloon part of it. That was the part of the story we were told to give to the public and news". DuBose also stated that there could be no doubt that the order to cover up the story came directly from General Ramey.

When Major Marcel returned to Roswell, he became a laughing stock. Everyone labelled him as the man who had misidentified a weather balloon as something out of this world. Yet when he was interviewed in 1978, he insisted that what was found on the Foster Ranch in July 1947 was definitely not a weather balloon.

Roswell remains one of the most important events in the history of UFO investigations. In the years since the event happened, countless witnesses have been interviewed, and hundreds of UFO researchers have attempted to uncover the truth.

The government has since declassified information that leads sceptics to believe that the weather balloon was actually part of a top secret experiment as part of

Project Mogul, designed to detect Russian nuclear tests. This remains disputed by those present in Roswell at the time of the incident, as well as those who have thoroughly researched the event. What cannot be disputed, is that something came down on Foster Ranch in July of 1947 and the military acted quickly to take control of the situation…

KECKSBURG, PENSYLVANIA
(9TH DECEMBER, 1965)

Across nine American states and as far north as Canada, thousands of witnesses reported seeing a large object streak across the sky on 9th December, 1965. Sources at the Pentagon would suggest that the object was nothing more than a meteor, but what happened in Kecksburg that night, would call that particular explanation into question.

The object was described as a "large fireball" and dropped hot metal over Michigan and Ontario which led to a number of large land fires over three states.

The Associated Press reported shockwaves and sonic booms along the trajectory of the object.

The object reportedly travelled to Greensburg, during which time shockwaves were reported in western Pennsylvania, before changing direction and ending its journey in Kecksburg.

One eyewitness account stated that as the object flew towards the mountains, it "seemed like it hesitated" before making a U-turn towards Kecksburg.

Further eyewitnesses stated that the object was "acorn-shaped" flying only around two-hundred feet above the ground. It appeared that the tip of the object was coated in some kind of vapor, and that the object itself was a greyish-brownish colour with smooth edges and smooth lines.

According to those that saw the object in detail, it definitely wasn't a meteor and seemed to be something that had been constructed as opposed to something that could have been naturally formed.

The object appeared to crash-land around 4:45 p.m. in the woods.

 As the press arrived at the scene, they reported that the whole area had been cordoned off by a large military presence that had taken control of the situation.

Anybody who tried to get close were prevented from doing so, and warned that there may be radiation.

The headlines the next day would read, "ARMY ROPES OFF AREA. 'UNIDENTIFIED FLYING OBJECT' FALLS NEAR KECKSBURG"

The first person on the scene was a man named Bill Bulebush. He claimed that at first when he entered the woods, he thought he was watching some youngsters playing with sparklers.

As he made his way closer, Mr. Bulebush stated that he

could see blue flames between the trees and heard a sizzling sound. When he shone his torch on the object, he said it was a burnt orange colour with blue flames coming from it, and was slightly embedded in the dirt nose-first.

The object had a raised area around it towards the back, which corroborates the earlier reports of the object being acorn-shaped.

According to Mr. Bulebush, there were what looked like "Egyptian markings", or in other words, hieroglyphics. Interestingly, the markings described at Kecksburg share similarities to those described as being present on debris collected at the Roswell incident some eighteen years earlier.

It wasn't long before the local fire department were alerted to the incident. The call came in requested a search team as it was believed that an aircraft had crashed in the woods. The call was taken by James Romansky, a volunteer firefighter working out of Latrobe. He had witnessed the fiery object coming from the north before the alarm was raised.

The fire department searched the area on foot and came

across the crash site. Mr. Romansky described the object as being large and metallic. However, unlike any conventional aircraft, there didn't appear to be any windows or doors, no method of propulsion such as a motor or engine were present, and it had no wings or aerodynamic features.

The object had no seams or rivet marks to show how it had been put together, and appeared to have been cast as it was perfectly smooth. Mr. Romansky's description of the site appeared to corroborate Mr. Bulebush's account as he also stated that the front end of the object seemed to have ploughed into the ground.

He explained that the object they had found in the woods was around twelve feet long and twenty-five feet in diameter. It had strange markings around the base which Mr. Romansky states were reminiscent of the ancient Egyptian hieroglyphics, a fact which also aligns with Mr. Bulebush's description. He explained that the object appeared to have a bumper around the base around six to ten inches wide that stood out, where the symbols were located.

Mere minutes after the fire department had located the

site, the military arrived and restricted the area. Mr. Romansky and his team were forced to leave.

Another person to witness the event was Jerry Betters. He described a watching a large army truck with a white star on the side driving out of the woods with an unusual object on the back of it. According to his account, it was copper-coloured with a domed top and hieroglyphics around the base.

A few days after the object had been recovered, a truck driver came forward with a sensational story. Simply known as "Myron", he claimed that his firm had received an order for sixty-five hundred double-glazed bricks to be delivered to Wright-Patterson Air Force Base in Ohio. The same base that the Roswell debris was allegedly transported to.

The order was for "building a double-walled shield around a recovered radioactive object".

He stated that whilst unloading the bricks, he noticed a "bell-shaped" object resting on stilts, surrounded by parachute-like sheets covering it. He managed to get a better look at the object through a small opening in one of the coverings.

The object he saw "looked like a large acorn around about fourteen feet to the top and about twelve feet wide at the bottom". He went on to explain that the object had a collar around the bottom with writing on. Myron's description of the craft he saw at Wright-Patterson is so similar to the object described at Kecksburg, and around the same time, that it would be difficult to believe this wasn't the same craft. However, Myron went on to state that he witnessed something even more sensational.

He claimed that in the same building as the object was a work bench around twelve feet long, which had a body lying on top of it under a white sheet.

He stated, "I saw the left hand of whatever was in there sticking out from underneath a sterile white pad and it only had three fingers". Myron's report went on to state that the body was between four feet five inches and four feet nine inches long, with dark greenish brown skin, like a lizard.

In 1990, a former Army Seargent, Clifford Stone, claimed that he had read official documentation relating to the Kecksburg object while still working for the military. He stated that the document detailed the recovery of an object in the Pennsylvania area that was not Soviet, and that the document "makes it quite clear that the object did not originate on the face of the Earth".

During the week of the crash, there were not only the Geminid meteor showers, but also a soviet satellite, known as Kosmos 96, that re-entered the atmosphere around the night of the Kecksburg incident.

This was confirmed by The Naval Space Surveillance Centre in Virginia. It is interesting that Sergeant Stone's testimony of the documents specifically state that the object was not Soviet.

What further fuels scepticism over the object being the re-entry of Kosmos 96, is that in 2003, Chief Scientist for Orbital Debris at NASA, Nicholas Johnson, was questioned about the object by an investigative Journalist. He stated, "I can tell you categorically, that there is no way that any debris from Cosmos 96 could have landed in Pennsylvania anywhere around 4:45 p.m. That's an absolute. Orbital Mechanics are very strict".

However, Sergeant Stone went on to say that he was informed by inside sources that Kosmos 96 had collided with an object of unknown origin that had in turn, caused it to re-enter the Earth's atmosphere. Whatever landed in Kecksburg appeared to be of extreme importance to the military.

According to Sergeant Stone, the object that fell from the sky that day was considered so sensitive, that those in the know would even lie to congress to keep the information "highly classified and buried away from the public".

If it were the remains of Kosmos 96, it would go some way to explaining the military presence at the site, though this seems a sheer impossibility based on the data from NASA. If it was simply a meteor, as officials suggest, then why would the military swarm the site in such a way?

Is it possible that the object was extra-terrestrial in nature?

As people delve deeper into the mystery at Kecksburg, all they seem to discover are more questions than answers...

39

MELBOURNE, VICTORIA
(6TH APRIL, 1966)

One of the strangest encounters of the last century took place in the small suburb of Clayton, Melbourne, at Westall High School where hundreds of people witnessed a UFO for over twenty minutes.

At around 11am, on the morning of 6th April, 1966, the school bell rang for morning recess. As the children walked out into the yard to enjoy their free time, they noticed girls from the previous physical education lesson, still wearing their kits, were gathered by the end of the playing field. As the children approached, they noticed something unusual in the sky.

One of the girls described looking up and witnessing a silver object flying around some pine trees behind the school. She claimed that the object flew over some open paddocks then back to the pines before descending into the small field at the rear of the school grounds. Shortly afterwards, the children watched as a number of small aircraft flew towards where they saw the object descend. When they reached the area of the pines, the object rose back up and hovered at the same altitude as the aircraft. The girl stated that it was as long as a Cessna airplane, but very thin.

As the aircraft began to approach the object, it tilted onto its side at a forty-five degree angle, and accelerated away too quickly for the aircraft to follow. The girl claims that after school, her and her two friends went to the field where the object had descended and found a spot where the long grass had been crushed into the earth. The area was around twenty-five feet in diameter.

The science teacher at Westall High School, Mr. Andrew Greenwood, gave his account of what he witnessed that day to the local press. He stated that there were five light aircraft circling the object at a low altitude.

He described the object as silver-grey and about half the length of the aircraft. The object remained in motion, moving up and down and side to side. One plane was observing the object at first, but by the end there were five which attempted to follow it as it accelerated back and forth.

This continued for around twenty minutes, at which point Mr. Greenwood turned away, and when he looked again, the object had disappeared.

Multiple pilots were later identified and spoken to, though none indicated that they had seen anything in the area at the time.

Another student described the object as "round, with a hump on top, and round things underneath".

The reports from the students were slightly varied. It could be that some of the students witnessed the object from one part of the school or at a different angle, and saw something additional to what was witness by a different group from a different part of the school or another angle.

It could also be that some of the students embellished details as children and some teenagers often do. However, the descriptions of the craft largely seem to be relatively consistent.

One of the students, known as only as "B", reported seeing three objects that day. According to her, two had landed in the field and burnt the grass in two large circles.

She estimated that the size of the objects were twice as big as reported by other students, at close to fifty feet in diameter. She also made a claim that the military became involved shortly afterwards.

According to her statement, military personnel had visited her house the following day where she lived with her mother and younger brother.

The family were escorted to the landing site, where soil samples were taken. One of the military personnel were overheard stating that the ground must have been subjected to extreme heat, as the earth itself had been burnt.

When they were returned to their home, she stated that the military had warned them not to speak about what they had witnessed to anybody. The girl's younger brother confirmed her story in part.

According to some sources, a few pupils at the school who spoke about what they had witnessed to Channel Nine had summarily been given detention.

Years later, a school prefect came forward to describe what he had witnessed. He described how he was in chemistry class at the time the object appeared and explained that it was a "classic flying saucer shape". According to some of the other students he spoke to around the time, there had been two objects, but he claims to have only witnessed one. He went on to state that the object he saw was silver in colour, and came down behind some pine trees around two hundred metres away.

What happened next, he described as "like a mass evacuation of the school". Teachers and students alike rushed out of the school to look at the object.

The prefect explained that behind the trees where the object had descended, they found a large area of flattened grass about thirty feet across. There were three scorch marks present and the flattened area formed a perfect circle.

A man approached the school group and angrily told them all to leave due to them being stood on private property. He then stormed off through the flattened patch of grass as though it didn't exist. The children tried to tell him a flying saucer had landed there, to which the man shouted obscenities at them before storming off.

The prefect claims to have got into a lot of trouble for being irresponsible in following the other children outside, and an assembly was held where the students were forbidden to talk to the press.

However, some students ignored this causing the principle of the school to make an announcement that no such thing had been witnessed, and claiming that everyone who said anything otherwise was suffering from mass hysteria.

The prefect went on to claim that he witnessed a couple of unmarked planes flying over the area at the time very slowly as they approached the landing site. Despite not seeing a second object himself, the prefect stated that there was a second area of flattened grass without scorch marks to the south. This suggests that there was a second object that landed, as B had reported previously.

According to the prefect, one of the science teachers had taken some photographs of the site and some of the students who ran towards the field claimed to have seen the object take off, though most only saw it come down behind the trees.

It is believed that the school staff were told by the military that the object was a top secret air force test.

One unnamed witness claims to have been at the site

around 5 p.m. on the day of the sightings and saw a "perfect circle of flattened grass" in an area he estimated to be around thirty feet in diameter.

A few days later when he returned to the site, he found a team of military personnel examining the area with what appeared to be radiation detecting equipment. The military told him he couldn't enter the field and had to leave.

The third time the man returned, the paddock and surrounding area had been burnt to a crisp.

Some other unusual events have been spoken about in recent years that were linked to the strange object at Westall, though these haven't been verified.

Rumours persist that the science teacher who took the photographs at the site was told that if he wanted to continue teaching he had to keep his mouth shut, and was forced to hand his film over to the military, and that the first girl to arrive at the site was found in a trance-like state and had to spend time in hospital as a result.

Something extraordinary happened that day, something that has never been fully explained.

It is possible that this was some form of experiment by the military, and that the children all witnessed something they shouldn't have, just as it is equally possible that there is an otherworldly explanation. One thing is for certain, that someone, somewhere, knows the truth...

48

SHAG HARBOUR, NOVA SCOTIA
(4TH OCTOBER, 1967)

In the southern tip of Nova Scotia is Shag Harbour, a small fishing community that witnessed one of the most unusual and thoroughly investigated encounters of the last century. The event itself is not as well-known as others, such as the incident in Roswell in 1947, but is equally as incredible.

On October 4th, 1967, a number of residents of Shag Harbour reported seeing an unusual group of orange lights in the evening sky. There were multiple witnesses that described seeing four lights above the water including a group of teenagers that claimed to have seen the lights flashing for a while before making a steep dive towards the surface.

According to those present, when the lights it the surface of the water, they didn't submerge straight away. Instead, they seemed to float on the surface around half a mile from the shoreline.

Originally, the witnesses believed they were watching an airplane performing an emergency crash-landing. The Royal Canadian Mounted Police in Barrington received the first report from a fisherman who described the incident as an airliner crashing into the bay. At first, they believed the young man had been drinking, but when more calls came in reporting the same incident, they realized something unusual had occurred.

Around the same time, RCMP Constable Ron Pound was on patrol, heading towards Shag Harbour. He had been watching the strange lights and sped up towards the area. Pound reported that he believed all the lights were part of a single aircraft that he guessed was around sixty feet long.

Officer Pound was joined by another two officers at the shore of the bay. There were several residents of the village stood along the bank watching the events unfold.

According to statements made by the officers, the lights began to slowly change from orange to yellow, moving across the surface of the water leaving behind a strange yellow foam.

Around thirty witnesses to the event would later testify that the object was around sixty feet long, around ten feet high, and dome shaped.

After watching the object for a few minutes, the

witnesses claim that it began to sink below the waves, some claiming they heard strange noise as it did so, likening it to the sound of wind rushing past them. Two of the officers and a few fisherman launched small boats toward the area where the object sank to search for any survivors. Those that did so, claimed that the lights were not visible when they reached the area. They also reported that they felt uncomfortable sailing through the thick, yellow foam, stating that it was like nothing they had ever seen before and was definitely not seafoam.

Several hours later, the search was called off at around 3 a.m. The search had yielded no results. Upon contacting NORAD and the Rescue Coordination Centre in Halifax, the RCMP were informed that there were no reports whatsoever of any civilian or military aircraft being lost or losing contact.

The following day, the RCC filed a report stating that something of "unknown origin" had crashed into the water in Shag Harbour. HMCS Granby was dispatched to Shag Harbour to conduct further investigation. Using some of the most advanced detection system they had, and sending down specially trained Navy divers, the Canadian military searched the bay for days looking for the object but reportedly found nothing. The story seemed to fade away as a simple historical oddity with no real explanation until, in 1993, two investigators from MUFON, the Mutual UFO Network, began interviewing witnesses in an attempt to discover the truth of what happened on that October night, back in 1967. They came across some previously unknown information that adds another layer of mystery to this compelling incident.

Divers and crew members from HMCS Granby had been interviewed, and they stated that the object had actually been tracked. According to their statements, it had travelled around twenty-five miles to a place called Government Point.

Government Point at the time of the incident contained a small military base operated by the United States, for the purpose of detecting and tracking submarines in the North Atlantic.

The United States military at Government Point had supposedly detected the object using their advanced detection equipment and dispatched a number of vessels to the location. These vessels positioned themselves over where the object had come to a stop. The object didn't move or react to their presence. Three days went by with no activity from the object, so the military began to plan a salvage operation. However, before they were able to launch such an operation, their equipment detected a second object approaching their location. It stopped close to the first object. According to those present, they believed the second object was offering assistance to the first.

For almost a week, the Navy remained in position, observing the situation and unsure of what they were witnessing. However, a Russian submarine had made its way into Canadian waters, and so a number of the Naval vessels had to head north to investigate.

As the vessels began to move, both objects accelerated underwater toward the United States. The remaining vessels pursued them as best they could, but the objects broke the surface of the water and flew into the sky before vanishing in mere seconds.

Many other witnesses corroborated these accounts, yet

this was not the official explanation given by the military.

Something undeniably occurred at Shag Harbour that was strange at best. It was enough to panic the residents of the small fishing village in Nova Scotia, and appeared to draw a lot of attention from the military.

Many people sceptical of the event claim that what came down was nothing more than a meteor. Those that witnessed the incident maintain that what they saw was under intelligent control. The truth is that we may never know.

Subsequent investigations have as yet failed to turn up any further evidence for either explanation that could finally put this mystery to bed…

55

RENDELSHAM FOREST, SUFFOLK (26TH DECEMBER, 1980)

An American air base in the United Kingdom, RAF Woodbridge, would be the site of one of the most high profile UFO cases the UK had ever encountered in the early hours of the morning on 26th December, 1980. Two of the base personnel, John Burroughs and Jim Peniston were on patrol when they claim to have witnessed a strange light in the skies over a nearby woodland situated next to the base. They believed that an aircraft, perhaps a helicopter, had crash-landed. They asked for permission to investigate the site and took Airman First Class Cabansag along for the ride in their jeep. However, on the way, the group encountered some rough terrain and had to abandon their vehicle, continuing the journey on foot.

The men reported witnessing a "strange glowing object" that was "metallic in appearance and triangular in shape", estimated to be around two to three metres across, around two metres high. The light from the object lit up the whole forest. On top of the craft, they would describe seeing a pulsating red light while a bank of blue lights sat underneath.

The intensity of light made it difficult at first to see

whether the object was hovering or had landed. The radios that the men carried began to experience some interference, which became more intense as they moved closer to the craft. Officer Penniston ordered AFC Cabansag back to the jeep in order to retain communication with the base.

When Officer Penniston and Officer Burroughs arrived around fifty meters from the site, the pair were astonished by what they saw.

Officer Burroughs was so focused on the object, he appeared to forget himself, whereas Officer Penniston was taking a mental note of everything, such as how the air felt so static that his hair began to stand on end. He continued to move closer to the object, to within twenty meters, and could see that the outside of the craft was black, and appeared to be smooth and shiny like glass. He would later describe how the blue and red lights would pulse alternately while there was a constant white light emitted from the top and bottom of the object.

It was at this point that Officer Penniston took out a camera and began to take multiple photographs of the object. He circled the craft while taking them, so as to get as much of the object photographed as possible. He claimed that he couldn't see any seams or rivets, as though the object was made from one single piece of material, and couldn't locate any visible signs of propulsion. Officer Penniston would then walk right up and place his hand on the surface of the craft, which he remembered felt warm.

Officer Penniston also described that there were raised etchings on the craft that felt like sandpaper. These etchings were of strange symbols that he stated resembled Egyptian-style hieroglyphics. They were around three inches tall, ands stretched across around three feet of the side of the craft.

When Officer Penniston touched the symbols, a bright like took over everything. He later stated that when he was surrounded by the light, he could "neither hear, nor see" and at one point, he thought the object was going to explode and backed away. The object then proceeded to rise into the air and moved away into the trees, while animals could be heard all around making frenzied sounds.

Even though Officer Penniston didn't witness any occupants in the craft, he claimed to have a strange feeling that something of higher intelligence was alive inside of it. While this whole situation unfolded, Officer Penniston continued making notes in his logbook. He still has possession of the logbook to this day, and notes that the speed of the craft was far beyond anything we could engineer.

Around an hour later, both Officers were back on base being debriefed. They were told that they should "forget they had ever seen anything".

Officer Penniston and Officer Burroughs believed that a cover-up operation was taking place, and as soon as daylight broke, they secretly made their way back to where they had come into contact with the object.

In the area where the object had been, the pair found three indentations in the ground that would suggest the object had landed as opposed to have been hovering. Officer Penniston claims to have taken plaster moulds of these indentations.

The story, however, takes a bizarre turn after he had returned to his room. According to statements made by

Officer Penniston, he began to have visions of what appeared to be binary code made up of sequences of 1's and 0's which he felt compelled to write down.

He took these codes to a binary expert who decoded them as reading, "EXPLORATION OF HUMANITY 666 8100 – CONTINUOUS FOR PLANETARY ADVANCEMENT".

It then listed a series of numbers, which were later determined to be co-ordinates.

The coordinates were identified as historically significant locations around the world, such as the Pyramids at Giza and the Nazca Lines in Peru among others in Greece, China, and elsewhere.

The following day, after Officer Penniston and Officer Burroughs return, Deputy Base Commander, Charles Halt, visited the site for himself. Upon arrival, his team detected increased levels of radiation in the area where the object had landed.

Interestingly, a nearby tree was irradiated only on the side that was facing the craft. During Commander Halt's visit to the base that evening, a second sighting occurred.

Commander Halt was issued an order to investigate the incident discreetly, though he would later express that his superiors wanted him to discredit the incident, no matter what his investigation uncovered.

As he investigated the area, he claimed to have witnessed a "glowing red object" speeding through the woodland. Commander Halt was carrying a tape recorder with him, and quickly hit the record button. The tape itself would be made public years later.

The men on the recording sounded genuinely anxious about what they were witnessing and during this encounter, the object rose above the trees, split into five separate white objects, and disappeared.

On the recording, Commander Halt can be heard giving a second by second commentary of what he was seeing in real time. He says, "It's coming this way. Pieces of it are shooting off. There's no doubt about it. This is weird. We are observing what appears to be a beam coming down to the ground. This is unreal."

As Commander Halt's team watched, a second object came into view and hovered above them. He claims that it also fired a beam into the ground that impacted just a few feet from where they were stood.

In the years since the event, it has been alleged that a high-ranking military official visited the base shortly after the incident and removed evidence pertaining to the object, though these reports have not been verified.

What is intriguing, is that those involved had to sign an oath not to disclose anything they had witnessed for a period of thirty years.

Why would the military want to keep such an event a secret?

If there is nothing to hide, then why all the secrecy?

Some sceptics will often claim that what people witnessed in 1980 was a secret military test, which may explain the secrecy.

Though the most commonly accepted theory put forward by those who believe that there is an earthly explanation for the events , is that the only thing the witnesses saw was a beam from a nearby lighthouse travelling through the trees.

Commander Halt, Officer Burroughs, and Officer Penniston all refute this explanation.

This encounter remains one of the most compelling

UFO events in the UK and to this day, many are still searching for the answers. It is rumoured that even now, nothing can grow on the area where the craft supposedly landed…

DAYTON, TEXAS
(29TH DECEMBER, 1980)

In late December of 1980, Vickie Landrum, Betty Cash, and Vickie's seven-year-old grandson Colby were travelling to Dayton, Texas, after going out to dinner. Around 9 p.m. while driving down a long stretch of road through the woods, they witnessed a light above the trees nearby, which they first thought to be an airplane.

A short time later, they saw another light but this seemed closer and brighter than the first. They believed it was the same light they had seen before until they noticed that the light was coming from a large, diamond-shaped object hovering around the same height as the trees. They claimed that they saw flames at the base of the object and could feel an intense heat. Mrs Landrum told Mrs Cash to stop the car. Mrs Cash thought about turning the car around but later stated that she felt they may get stuck as the road was quite narrow. It had been raining, and she didn't want to get the trio into any difficulty in the soft dirt surrounding them.

The two women got out of the car, but because her grandson was so scared of what he was witnessing,

Mrs Landrum quickly returned to offer him comfort. Mrs Cash stayed outside, watching the object as if mesmerised by it. The object was described as being intensely bright, metallic silver in colour, and shaped like an upright diamond about the size of a water tower, with its top and bottom flattened rather than pointed.

There were small blue lights in a ring around the centre of the object. Flames flew out of the bottom periodically, flaring outwardly creating a cone shape. Whenever the flame disappeared, the object lowered a short way before they would kick in again and the object would ascend slightly, much like the operation of a hot air balloon.

The heat was so powerful, that it made the body of the car painful to touch. Mrs Cash claimed that she had to use her coat to open the door, in order to protect her hand from being burnt. When she got back into the car, she touched the dashboard which had become so soft that her hand left an imprint that remained present for weeks.

Shortly after Mrs Cash returned to the car, the object

rose up above the trees. A group of black helicopters approached in formation and surrounded the object. According to both Mrs Cash and Mrs Landrum, there were twenty-three helicopters in total.

Some of them were identified as having two rotors, which many believe were CH-47 Chinook helicopters which are used primarily by military forces throughout the world. At that point, Mrs Cash began to drive, still keeping her eye on the object which disappeared into the distance.

The whole incident lasted around twenty minutes. However, this wasn't the end of the encounter for the trio, as all started experience ill health later that evening. Along with Mrs Cash, the Landrum's both suffered from weakness, vomiting, diarrhoea, burning in their eyes and a feeling they described as similar to sunburn. Unfortunately, Mrs Cash appeared to suffer from more severe symptoms.

Her skin formed many large, painful blisters and she was admitted to hospital on 3rd January, 1981. She could not walk at the hospital and had lost clumps of her hair. It was also reported that patches of her skin had fallen off.

After twelve days in the hospital, she was released, but returned to hospital for another fifteen days when her symptoms didn't improve. She later stated she was given cancer treatment due to radiation emitted from the object. Both the Landrums would later suffer from skin sores and hair loss, though to a lesser degree than Mrs Cash.

During a MUFON investigation, a radiologist examined the medical records of those involved and claimed, "We have strong evidence that these patients have suffered secondary damage from ionizing radiation. It is also possible that there was an infrared component as well".

However, the speed at which the symptoms manifest would be only consistent with such a high dosage of radiation, the victims would have died within a few days, which has led some to believe that the symptoms were more likely caused by the witnesses coming into contact with some kind of chemical, possibly in aerosol form.

A police officer was later identified by investigators

who claimed to have seen twelve Chinook helicopters close to the area where the incident occurred, but stated that he did not see the object.

What is interesting about this event, is that Mrs Cash and Mrs Landrum sought financial damages for their injuries through the courts. Officials from all branches of the military, and NASA, gave evidence in the case, until the case was dismissed by a District Court Judge. The reason given for dismissal of the case, was that it could not be proven that the helicopters belonged to the United States government. Additionally, since the military officials testified that they did not have an aircraft matching the object's description in their possession, both women lost their case.

What is clear about this case, is that there can be no doubt Mrs Cash was admitted to hospital with severe injuries. It could well be that these were caused by the object they witnessed along that stretch of road, and without any concrete explanation for what occurred that night, the event remains a mystery…

70

BRUSSELS, BELGIUM
(30TH MARCH, 1990)

From October, 1989, through to April of 1990, thousands of witnesses reported seeing unusual objects in the skies over Belgium. These sightings began near Eupen and were mostly witnessed at night, although some were reported during daylight hours.

Many of the witnesses were officials in one capacity or another from the police and the military. The most prominent description of the objects were that they were triangular and black, with lights on each corner and a flashing, red light located in the middle of the craft. Some reports also claimed that there were more lights on the sides of the craft.

The size of the craft was reported by some as being the size of a football field, travelling very high, while others claim it was at a low altitude flying overhead.

While these discrepancies allow some sceptics to discredit the witnesses, it is worth noting that these reports were made over a prolonged period of time, and so it could be possible that people were seeing different objects.

Some of the craft reportedly flew at very slow speeds and would often hover in place, while other sightings described how the objects would fly from one point on the horizon to another point directly opposite in just seconds.

While most of the reported sightings claimed that the objects were noiseless, some of the craft appeared to emit a low level humming sound.

In a number of the lower altitude sightings, witnesses claimed to see heavy-looking metal parts and tubes present on the underside of the craft, along with criss-cross effects and diamond shaped objects.

Some of the other reports claimed that the craft were transformative, with some witnesses stating that they watched as one of the triangular shaped craft reconstructed itself into a circular shape with a light underneath. It then released a large amount of small, red objects that spread out in all directions before the object reverted back to its original triangular form.

Two police officers witnessed an object emit a bright light onto their car. They described it as being a black triangular craft that had three lights on the underside. One in each corner. It also had a red flashing light in the centre, and emitted a low-level humming sound.

The craft was very large and flew at an altitude estimated to be between six hundred and nine hundred feet above the ground. According to other witnesses, the craft then travelled slowly towards the Gileppe Dam where it hovered for three quarters of an hour before then heading to the city of Spa where it sat for thirty minutes before disappearing.

Another member of the police force was at the Eupen station saw a rectangular craft glowing brightly outside. He estimated it to be around sixty-five feet in length. He watched as the craft moved away and disappeared, then contacted two police officers who were in an area near to where the object was heading. Between five and ten minutes later, the officers spotted the craft. They contacted SOPEBS (Société belge d'étude des phénomènes spatiaux) which is the Belgian Society for the Study of Space Phenomena and reported the object to them. They could see the underside of the craft and stated that it was triangular shaped. As it moved, they caught a glimpse of a dome located on top.

This series of sightings became known as The Belgian Wave. However, the most interesting case took place towards the end of the wave, on 30th March, 1990 at around 11 p.m. On that evening, unknown aerial phenomena were tracked on radar, and sighted by an estimated thirteen and a half thousand people, of whom, two thousand six hundred witnesses filed official reports.

Strangely, rather than disregard the reports or cover up the events, the Belgian Air Force were surprisingly open with the public, releasing their own report into the incident.

At around 11 p.m. the supervisor for the CRC (Control Reporting Centre) began receiving reports that three unusual lights were seen heading in the direction of Thorembais-Glons, to the south-east of Brussels. The lights were described as being much brighter than stars and changing colour between green, yellow, and red. They appeared to be in a triangle formation.

The CRC took the step of contacting the local police to confirm the sighting.

Around ten to fifteen minutes later, witnesses described seeing a second set of lights heading in the same direction as the first.

By 11:30 p.m. the police had confirmed the initial sighting and the objects had appeared on radar at the CRC. The second set of lights had been moving erratically and appeared to have no structure, but had by this point formed themselves into a smaller triangle formation.

Shortly before midnight, the CRC had received second radar confirmation from Traffic Centre Control at Semmerzake Glons and issued an order to scramble two F-16 fighter jets from Beauvechain Air Base.

The objects were still clearly visible from the ground. It was reported that during this time, they moved slowly across the sky while maintaining their formation. During the next hour or so, the two F-16s attempted to intercept the targets nine separate times. They managed to lock on for a few seconds on three different occasions but each time, the object managed to outmanoeuvre the jet at high speed, breaking the lock.

According to the report, during the first radar lock, the target made some incredible manoeuvres. It accelerated from 240 kilometres per hour to over 1,700 kilometres per hour. During this time, the object descended over nine hundred metres to almost ground level in less than two seconds.

Similar movement was observed during both the second and third radar locks. At no point did the pilots make visual contact with the objects, and at no point did anyone report hearing a sonic boom. The sudden changes in speed and altitude would have been fatal for a human pilot, even if we had the technology to manoeuvre in such a way.

Witnesses on the ground seemed to corroborate the story, as they described seeing the object moving very rapidly with the F-16s in pursuit. The final confirmed radar lock took place at 12:40 a.m. but was broken by the object accelerating from around 160 kilometres per hour to around 1,100 kilometres per hour.

After this lock was broken, all contact was lost. The F-16s returned to the base shortly after 1 a.m. due to running low on fuel.

The Belgian Wave was one of the most important European UFO events in history. However, a few years later, in 1992, Marc Hallet, a Belgian sceptic, wrote an essay claiming that the Belgian UFO Wave was nothing more than a mass delusion, while Renaud Leclet claimed in his hypothesis that the sightings could be explained as being helicopters, and the lack of noise could be due to engine noises from the witnesses' vehicles or a strong natural wind.

The argument over what was witnessed by so many people rages on to this day, with no real explanation yet to come.

What makes this event so important, however, was the open and honest way in which the Belgian military dealt with the event. It could be said, that their actions were responsible for the public remaining calm and not becoming panicked during this unusual encounter. Many UFO researchers point to the Belgian Wave as being strong evidence supporting claims of visitation by an extra-terrestrial presence...

PHOENIX, ARIZONA
(13TH MARCH, 1997)

Something extraordinary flew over Phoenix, Arizona, in 1997 which has left many questions in its wake, with no concrete answers. On March 13th, something silently crossed the skies above the city leading to the event being widely regarded as one of the most documented UFO events in history.

That evening, a large number of witnesses reported seeing unusual lights in the sky overhead. The first report recorded was from a man in Nevada, who claimed to see a V-shaped object the size of a Boeing 747 at around 7:55 p.m. He stated that the object had six lights along its edge and made a sound like "rushing wind". It was heading towards Phoenix.

The second sighting reported was around 8:15 p.m. by a former police officer from Paulden. He claimed to have been driving when he spotted five "reddish-orange" lights to the northwest. He stated that four were lined up together, while one was lagging behind. Within minutes of the second sighting, national UFO reporting centres were flooded with calls from within both Nevada and Arizona. A large number of calls were also made to local police departments, the press, and nearby Luke Air Force Base.

The calls continued for many hours, with the majority of them reporting a solid structured craft that appeared to be so large it would block out parts of the night sky. John Kaiser, a key witness to the events, reported that his family noticed a group of lights to the north west of their home in Prescott. He stated that the lights were all red except for a leading light on the nose, which was white.

The lights were arranged in a triangle formation, and Kaiser watched it pass directly overhead before it appeared to bank to the right and disappear to the south east. Kaiser was unable to state how high above him the object travelled, but stated that the object was completely silent.

Many other witnesses from Prescott Valley made reports that night of similar lights in the sky. Some people reported that the lights were yellow or off-white instead of red or orange.

It could be that the lights appeared differently to people due to atmospherics or glare, or it could be that the lights changed colour. What remained consistent throughout the reports, however, is that the craft made no noise whatsoever as it flew overhead, and that the shape of the craft was described as V-shaped, Triangular, or wedge-shaped.

The object continued on its path and reports started to come in from the Phoenix area. One report came from a woman who was driving just north of the city. She described the object as the shape of "sergeant's stripes" and stated that it blocked out a large amount of the sky and appeared to hover above her car for around five minutes.

According to the woman, the object fired a beam of light down to the ground, causing the other lights to dim slightly. After firing the beam, the object continued on towards Phoenix.

Another reported sighting of the object took place shortly afterwards of five distinct lights in an arc shape. The lights were heading towards the witnesses' home, moving very slowly.

They reported that as it got closer, they realised it wasn't an arc, but more of a V-shape and only seemed to be a hundred or so metres above them. They also reported that the object was silent as it headed towards Tucson before it disappeared.

A local truck driver reported seeing a second set of lights that evening, which were completely different from the V-shaped object reported by others. He stated that as he was driving along Route 17, he witnessed two bright lights hovering around one or two miles ahead of him.

Three fighter jets were scrambled from the nearby air force base and approached the objects to intercept them. When the jets got close to the first object, it flew upwards at an incredible speed as the jets passed by. Around 10 p.m. nine lights were witnessed appearing one by one behind the Sierra Estella Mountain Range. Thousands of people saw these lights fade out after several minutes, with many taking photographs and video footage of the event. Experts estimated the size of this craft to be up to a mile long.

The military was made aware of the object and asked if they had anything in the sky that night that could account for the unusual event. Their initial response was that they had no aircraft in the area. Strangely, the military would change their story and issue an official response to the event.

They reported that the lights were flares dropped from an A10 Warthog as part of a training exercise over the Barry Goldwater Range.

While this explanation could potentially explain the second sighting, it does not explain the many witness reports from earlier in the evening who reported seeing a large black object blocking out parts of the sky.

The Phoenix Lights, as they have become known, are continuously under investigation by various UFO groups and enthusiasts, even now.

Perhaps as our own technology advances, those people who are hard at work studying the various videos and photographs of the event may finally be able to provide some answers.

Until then, all we can do is wait…

84

CHICAGO, ILLINOIS (7TH NOVEMBER, 2006)

One of the most well-known UFO events in recent memory took place at O'Hare airport in Chicago and was witnessed by both passengers and airport staff alike. There have been multiple photographs and videos released of the event which shows something unusual hovering in the skies above the runway.

At 4:15 p.m. on November 7th, 2006, a number of witnesses from United Airlines reported seeing a strange object above Concourse C of the United Airlines terminal. It was described as being of typical flying saucer shape. The object was dark grey in colour and estimated to be between six and twenty four feet across.

Some people reported that the object was spinning somewhat like a frisbee or a spinning top, while others claimed it showed no movement at all.

One thing that all the witnesses agreed upon, was that the object simply appeared just below the clouds and hovered silently. According to those present, the object remained in position for some time before rapidly accelerating into the clouds so quickly that it left a round hole.

One mechanic was quoted as saying, "What I saw stood out very clearly, and it was definitely not an aircraft". One of the managers at Concourse C heard about the sighting and rushed out of his office to see it for himself. He claimed that he needed to identify the object because of its close proximity to their flight operations.

If someone was messing around with a balloon or something over the airport, they needed to be stopped due to the risk they posed.

Many major news outlets took an interest in the events at O'Hare and questioned what the object could be. Hoping for some insight into the strange craft, many were left disappointed when the Federal Aviation Administration refused to investigate the incident.

At first, both the FAA and United Airlines stated that they didn't know anything about what happened, but recorded calls and other evidence that has been made public since, shows that not only did they know about it, but they were also talking in detail about it at the time the encounter was occurring.

According to some, United Airlines told its employees

that they weren't to speak about the event.

The FAA later claimed that the event was meteorological in nature. An FAA spokesman, Tony Molinaro said that the event couldn't happen due to there being no evidence of the event. He claimed there had been nothing detected on radar.

An astronomer at the Adler Planetarium seemed to agree with their statement, reporting that the weather at O'Hare that day was perfect for the appearance of what is known as a "hole-punch" cloud.

The astronomer explained that it happens when the temperature is around freezing and a plane flies through a uniform cloud causing a disc of ice crystals to fall from the cloud looking like someone punched a hole right through them.

However, those that witnessed the event first hand, and the millions of people around the world who have since watched the videos or seen the photographs of the event on social media disagree entirely with that explanation.

Many people are concerned by the lack of interest the FAA showed, since this event took place just five years after the tragic events at the World Trade Centre. It seems strange that they wouldn't be concerned about something in the skies above the airport.

The National Aviation Reporting Centre on Anomalous Phenomena (NARCAP) compiled a 115 page report on the event at O'Hare and called upon the government to not only investigate, but also to look into better energy sensing technologies. They stated, "If an object can hover for several minutes over a busy airport without being picked up on radar or seen from the control tower, it could be dangerous for flights". Some experts claim that radar can't always find objects that don't look like conventional aircraft, or that move in unusual ways. They also struggle to pick up objects that move very quickly or remain stationary for long periods.

One expert in particular, John Callahan, was in charge of accidents and investigations for the FAA during the 1980s. He was reported as saying he was not surprised at all that the object at O'Hare wasn't picked up by radar, but that doesn't mean there wasn't anything there.

The events at O'Hare airport are one of the more recent mass sightings, but since there appears to be a shroud of secrecy surrounding the events, the incident currently remains a mystery...

THE FUTURE

When discussing these events, there's a lot that is open to interpretation. Those that believe the events are extra-terrestrial in nature have an uphill struggle to convince the world, while sceptics continue to claim that most of these events were caused by strange weather, meteors, mass delusion or were simply hoaxes designed to trick an unsuspecting public.

Across these cases in particular, there appear to be some common elements. Triangular shaped craft appeared in Belgium, Phoenix, and Rendelsham. Hieroglyphic symbols were reported at Aurora, Roswell, Rendlesham, and Kecksburg.

These could be coincidence. It could be that the symbols were simply misunderstood and could have been Chinese or Russian, rather than otherworldly. It could be that the triangle shapes witnessed by so many were some form of misidentified test aircraft that the military would rather keep under wraps.

Who can say for certain when the witnesses in these cases are adamant that what they saw and what they experienced was definitely not of this world?

It is hard to believe that with the large military

presence at most of these sites, the government wouldn't be able to provide any answers.

That is, unless they didn't want to.

Further events have given many UFO enthusiasts reason to believe that the government are hiding the truth, from leaked documents to former operatives coming forward with deathbed confessions, the future will continue to offer up evidence one way or the other.

In 1989, a man named Bob Lazar claimed to have worked for the United States Government at a place known as S4 which he described as a top secret facility near to Papoose Lakebed within the borders of Area 51. Lazar states that the government had recovered nine extra-terrestrial craft, and that they were being reverse engineered at the facility. His job was to work on the extra-terrestrial propulsion systems.

Interestingly, Lazar went into incredible detail about how these craft achieved flight, even explaining that they used an element known as Element-115. An element that didn't exist on the periodic table.

When Lazar made these claims, people quickly jumped to discredit him.

The University where he obtained his degree claimed he was never a student, his former employer at the Los Alamos Physics Facility where he claimed to have worked, stated that he had never been employed at the facility. However, Lazar has remained consistent with his story.

Even more fascinating, is that Element-115 was successfully synthesised in 2003, and officially recognised as a new element, Moscovium, in 2015. Twenty-six years after Lazar came forward.

Furthermore, the existence of Area 51 as detailed by Lazar, was constantly denied by the United States Government, until 2013 when it was officially acknowledged.

More recently in 2011, the Ministry of Defence in the United Kingdom released thousands of files relating to UFO reports they investigated between 1985 and 2007. Between 2007 and 2012, the Pentagon ran the Advanced Aviation Threat Identification Program. This was a top secret program designed to investigate the UFO phenomena.

A former intelligence official first claimed the project

existed in 2017, and in April of 2022, the Pentagon released over one thousand, five hundred pages of documentation from the program.

These included details of research into the biological effects of UFO sightings on humans, such as unaccountable pregnancies, along with studies into advanced technologies and the categorisations of paranormal encounters.

Many UFO enthusiasts and researchers believe that we could be close to the moment of disclosure, when the governments of the world finally admit that extra-terrestrials exist.

According to a 2021 study, around sixty-five percent of people now believe that extra-terrestrial life exists, and eighty-seven percent don't believe that UFOs pose a threat while a poll conducted in 1997 resulted in ninety percent of Americans believing that the government are hiding the truth about extra-terrestrials.

With more and more people in possession of camera-equipped smart phones, there has been a huge increase in reported UFO sightings, most of which nowadays make it into the public domain via social media platforms such as YouTube.

Could we be reaching a tipping point where it becomes almost impossible for the governments of the world to hide the truth any longer?

At what point, if at all, will the public be ready for disclosure? Or is it that there is nothing to disclose, and we are all seeking answers for the unexplained, that simply don't exist?

The future will undoubtedly bring the truth with it, whatever that may be, but until then all we can do is wait, watch, and wonder...

ABOUT THE AUTHOR

Phoenix W. Thompson grew up fascinated by the subject of UFOs and the possibility that extra-terrestrial intelligence may exist somewhere in our universe. He has spent a lot of time researching the subject as a hobby and enjoys keeping up to date with the latest developments.

After obtaining a BA(Hons) in Creative Writing and Film Studies at DMU, he released the first book in his science fiction series, The S4 Saga. He then obtained a Diploma in Ufology from the Centre for Excellence, before undertaking work on his MA in Creative Writing where he completed the final book in his S4 series.

Now both a member of MENSA and MUFON, Phoenix spends his time taking care of his children, all the while looking to the stars and asking the biggest question of all…

What if…?

www.ingramcontent.com/pod-product-compliance
Lightning Source LLC
Chambersburg PA
CBHW071538150726
48000CB00002B/850